Contents

1 Chapter 1: Ancient Grains

Many ancient grains, once staples in diverse cultures, have fallen into obscurity. These include amaranth, with its high protein and lysine content, a crucial amino acid often lacking in other grains; quinoa, a complete protein source adaptable to harsh conditions; and teff, a tiny grain exceptionally rich in iron and calcium, thriving in arid environments. Their resurgence is vital, offering nutritional density and resilience against climate change's unpredictable pressures. For beginners, sourcing these grains may require seeking out specialty stores or online retailers, initially.

Their remarkable adaptability sets ancient grains apart. Unlike many modern hybrids, they often exhibit innate pest and disease resistance, needing less chemical intervention. This intrinsic resilience is especially critical as climate change intensifies, bringing unpredictable weather patterns and increased pest pressures. Cultivating these grains reduces reliance on resource-intensive, environmentally damaging farming practices. We should focus on supporting local farmers and seed banks who preserve and promote these valuable crops.

Consider the nutritional benefits. Ancient grains frequently boast higher levels of fiber, vitamins, and minerals compared to modern counterparts. This enhanced nutritional profile contributes to improved gut health, reduced risk of chronic diseases, and overall better well-being. Their role in combating malnutrition, particularly in regions with limited access to diverse foods, is undeniably significant. They represent a path towards healthier communities and improved human capital, a crucial element of sustainable development.

The cultivation of ancient grains presents a valuable opportunity for small-scale farmers. Their lower input requirements, relative to modern high-yield varieties, make them accessible to farmers with limited resources. Furthermore, the potential for niche markets and premium pricing can significantly enhance profitability, making this a viable pathway towards economic empowerment in many communities globally. This is particularly true in areas where diverse, resilient crops offer competitive advantages.

Preserving biodiversity is paramount. Ancient grains represent a genetic treasure trove, containing unique traits that can be invaluable in breeding programs aimed at developing climate-resilient crops. The loss of these varieties would be an irreversible blow to agricultural diversity, reducing our capacity to adapt to future challenges. By actively integrating these grains into farming systems, we safeguard not only food security but also the rich tapestry of agricultural heritage. It is a critical issue for future food security.

Initiating a small-scale cultivation project begins with careful selection of suitable varieties. This includes researching grains that thrive in your local climate and soil conditions. Begin with a small plot, learning from your experiences and expanding gradually. Joining local farming communities or connecting with experienced growers provides invaluable guidance, overcoming initial hurdles and providing necessary support networks. Experimentation is key to discovery and success.

Processing and preparing ancient grains can differ from those of common wheat or rice. Quinoa, for example, requires rinsing to remove saponins, naturally occurring compounds that can impart a bitter taste. Many ancient grains cook similarly to rice or other grains, but understanding the specific cooking times and methods is essential to ensure optimal palatability and nutritional value. Learning from others' experiences is invaluable.

Integrating ancient grains into one's diet is straightforward. They can be incorporated into existing meals, used as a base for porridge or added to salads, soups, and stews. Experimenting with different recipes unlocks a world of culinary possibilities. From hearty grain bowls to nutritious baked goods, the versatility of ancient grains offers exciting opportunities for creativity in the kitchen. It's a journey of gastronomic discovery.

Remember, the journey towards incorporating ancient grains into your diet and farming practices is a progressive one. Start small, adapt to challenges, and celebrate successes. Sharing knowledge and experiences within your community empowers collective learning, accelerating the adoption of these valuable crops for long-term food security. Knowledge sharing is the cornerstone of resilient communities.

The benefits extend beyond the nutritional and economic. Cultivating ancient grains connects us to our agricultural heritage, fostering a deeper appreciation for the land and its resources. It encourages a more mindful relationship with food, appreciating the effort and resources that nourish us daily. Connecting to the land and its history is invaluable, especially in urban settings. This offers a profound sense of accomplishment and contributes to overall well-being.

1.1 Forgotten Crops

Forgotten crops, often overlooked in modern agriculture, represent a treasure trove of genetic diversity and resilience. These are plants once widely cultivated but now relegated to the fringes, their cultivation practices lost to time or overshadowed by the high-yield monocultures dominating industrial farming. Their revival is crucial not only for diversifying our food sources but also for building more sustainable and climate-resilient agricultural systems. Many possess inherent resistance to pests and diseases, reducing the need for harmful chemical interventions, a significant benefit for both environmental health and human wellbeing. Consider amaranth, a nutritious grain with a history stretching back millennia, possessing remarkable drought tolerance and high protein content. Or quinoa, a complete protein source thriving in harsh conditions, offering significant potential for food security in arid regions. These are not mere historical curiosities; they are vital tools in creating a more resilient food future.

The rediscovery and reintroduction of these forgotten crops demands a multifaceted approach. It necessitates dedicated research into optimal cultivation methods adapted to modern contexts, recognizing that traditional knowledge holds invaluable insights alongside contemporary scientific advancements. We need to actively preserve the genetic diversity within these plant varieties, safeguarding against genetic erosion through seed-saving initiatives and the creation of seed banks dedicated to heirloom varieties. This work should not be confined to the laboratory; it necessitates the involvement of local communities, particularly indigenous groups who have traditionally cultivated these plants and possess unique knowledge of their propagation, harvesting, and culinary uses. Their participation is essential for ensuring the cultural continuity and ecological integrity of these efforts. Furthermore, supporting the cultivation of forgotten crops requires a shift in agricultural policy. Current subsidies often favor large-scale monoculture farming, implicitly disadvantaging small-scale farmers who often champion biodiversity and traditional farming practices. Re-evaluating these policies to incentivize the production and consumption of underutilized crops is vital. Financial support, targeted education initiatives, and the development of robust market infrastructure – including processing and distribution networks – are all critical components for empowering farmers and encouraging widespread adoption. This transition will not be immediate, requiring patience, collaboration, and a long-term commitment.

The benefits extend far beyond mere food production. These crops often exhibit superior nutritional profiles compared to their modern counterparts, offering a rich source of vitamins, minerals, and essential amino acids. Their cultivation can contribute to improved soil health through diverse root systems and nitrogen fixation, enhancing soil fertility and reducing reliance on synthetic fertilizers. Many are well-suited to marginal lands, helping to reclaim degraded areas and promote sustainable land management practices. Moreover, the integration of these crops into existing farming systems can enhance biodiversity, creating more resilient and robust ecosystems less susceptible to shocks and stresses. The potential economic benefits are also significant, offering opportunities for rural development and income generation through niche markets and value-added products.

Successfully integrating forgotten crops into our food systems necessitates a concerted and collaborative effort. It requires the involvement of researchers, policymakers, farmers, and consumers alike. Educational outreach is crucial, raising public awareness of the nutritional benefits and environmental advantages of these underutilized crops. Consumer demand plays a vital role; actively choosing these options when available signals a market opportunity that incentivizes farmers and encourages further investment in this critical area of agricultural sustainability. The challenge lies in bridging the gap between traditional knowledge and modern agricultural practices, recognizing the value of both and leveraging their synergy to create more resilient and sustainable food systems capable of meeting the needs of future generations. The revival of forgotten crops is not simply about recovering lost foods; it is about building a more resilient, equitable, and environmentally sustainable food future. It's a challenge worth embracing for the health of both our people and the planet.

1.2 Resilient Varieties

Spelt, emmer, and einkorn—these ancient grains, often overlooked in modern agriculture, possess remarkable resilience. Their ability to thrive in harsh conditions, demanding less water and fertilizer than their modern counterparts, makes them invaluable assets in a world grappling with climate change and resource scarcity. We should remember that their nutritional profiles are often superior, boasting higher levels of essential minerals and fiber. Cultivating these grains isn't just about reviving historical practices; it's about securing a sustainable future for our food systems. Their adaptability ensures yields even under challenging environmental circumstances, offering a crucial safety net against unpredictable weather patterns and resource limitations. Investing in the research and development of these ancient grains is an investment in global food security.

Many resilient varieties boast exceptional disease resistance, requiring minimal pesticide use. This inherent defense mechanism reduces reliance on synthetic chemicals, promoting a healthier environment and safer food production. Furthermore, these crops often possess shorter growing seasons, making them ideal for regions with limited growing windows or unpredictable weather patterns. Their adaptability to varied soil conditions further expands their cultivation potential, minimizing the need for intensive soil management techniques. This inherent robustness contributes significantly to reducing overall production costs while simultaneously ensuring a reliable food supply. Consider the long-term benefits of decreased reliance on external inputs, a benefit that extends far beyond the farm itself.

The benefits extend beyond simple agricultural efficiency. These resilient varieties can contribute to a greater diversity of crops in our food systems. Biodiversity offers a buffer against disease outbreaks and climate shocks. A reliance on a limited number of high-yield varieties leaves our food systems extremely vulnerable. Diverse crop portfolios are more robust, less vulnerable to catastrophic failures. The revival of ancient grains strengthens this necessary diversity, safeguarding our future food security. This diversification offers not just resilience, but also a wider range of nutritional options for consumers.

However, the transition to widespread cultivation of these grains requires careful consideration. Their lower yields compared to modern high-yield varieties may necessitate adjustments in farming practices and marketing strategies. Educational programs focusing on growing and processing these grains are vital. Farmers need comprehensive support to understand their specific needs and optimize their cultivation methods for maximal yields and optimal nutritional value. It's also important to consider market demand, which might require targeted marketing campaigns to raise consumer awareness of these beneficial and versatile crops. Collaboration between farmers, researchers, and policymakers is paramount for overcoming these challenges and successfully integrating these resilient varieties into our food systems.

But the resurgence of ancient grains isn't solely a matter of practical agriculture; it also connects us to a richer history of food production. These grains have sustained human populations for millennia, serving as testaments to human ingenuity and agricultural adaptation. They represent a direct link to our past, a connection to traditional farming methods often proven more sustainable and environmentally sound than contemporary practices. Their use promotes cultural preservation and enhances food heritage, offering a diverse and flavorful culinary landscape. Moreover, their incorporation in modern diets contributes to a deeper appreciation for the history of food and agriculture. Rediscovering these grains is a journey of rediscovering our own food heritage.

The challenge lies in effectively communicating these benefits to both farmers and consumers. Many are unaware of the inherent advantages of ancient grains, and the transition will require educational outreach and support systems to help farmers adopt these resilient crops. Government policies can play a vital role in incentivizing their cultivation, perhaps through subsidies or other financial support. This might involve setting aside specific areas for their cultivation or supporting research initiatives to further improve their yields and adaptability. The collaborative effort is crucial for creating a sustainable food system that incorporates the best of traditional wisdom and modern agricultural advancements. Success demands a multifaceted approach.

2 Chapter 2: Soil Regeneration

Healthy soil is the foundation of any thriving agricultural system. Ignoring its importance is akin to building a house on sand; eventually, everything crumbles. Regenerative practices, unlike exploitative ones, focus on building soil health rather than depleting it. This leads to a more resilient, productive, and environmentally sound food system.

Composting, a cornerstone of soil regeneration, isn't just about waste disposal; it's about creating a nutrient-rich amendment that significantly boosts soil fertility. A well-managed compost pile transforms kitchen scraps, yard waste, and other organic materials into "black gold," a living soil conditioner teeming with beneficial microbes. This process is more than simply decomposition; it's the creation of a complex ecosystem that improves soil structure, aeration, and water retention—all critical factors for optimal plant growth. Learning proper composting techniques, from creating the optimal carbon-to-nitrogen ratio to managing moisture levels and turning the pile regularly, is an investment in long-term soil health and sustainable agriculture. Careful attention to detail yields rich, dark compost, a boon to any garden or field.

No-till farming offers a revolutionary approach, dramatically reducing soil disturbance compared to conventional plowing. The benefits are profound: increased organic matter retention, improved soil structure, reduced erosion, and less water runoff. By leaving crop residues on the soil surface, no-till farming creates a protective mulch that shields the soil from the sun's harsh rays, preventing moisture loss and maintaining a consistent soil temperature. This method also benefits the soil's biology by promoting a vibrant community of beneficial organisms that improve nutrient cycling and overall soil health. The reduction in machinery use also saves time, energy, and ultimately, money—a win-win situation for both the farmer and the environment. Transitioning to no-till might require patience and careful planning, but the long-term rewards are undeniable.

Cover cropping is a simple yet incredibly effective technique that involves planting specific crops to protect and improve the soil between cash crop plantings. These cover crops aren't harvested for food; instead, they serve as a vital part of the soil regeneration process. Legumes, such as clover and alfalfa, fix nitrogen from the atmosphere, enriching the soil with a crucial nutrient for plant growth. Other cover crops, like rye and oats, act as living mulch, suppressing weeds, improving soil structure, and preventing erosion. The decomposition of cover crops after being planted adds organic matter to the soil, enhancing its fertility and water-holding capacity. Planning ahead to choose the right cover crop for your specific conditions and soil type is essential for maximizing the benefits. Integrating cover cropping into your farming practices can be relatively easy, even for small-scale farmers, with significant rewards.

Understanding the crucial role soil plays in food production is paramount. Healthy, living soil is more than just dirt; it's a complex ecosystem teeming with life, a vibrant web of organisms working together to create a fertile environment for plant growth. Approaches focusing on soil regeneration are not only environmentally responsible but also economically advantageous. They lead to increased yields, reduced input costs, and greater resilience in the face of climate change. By embracing these techniques, we can move toward a more sustainable, food-secure future. The transition might require a shift in mindset and some initial investment in time and learning, but the long-term benefits are immeasurable for both the planet and our own well-being. A commitment to soil regeneration is a commitment to a healthier, more sustainable future. Investing in soil regeneration is an investment in a more secure future for everyone.

2.1 Composting Methods

Composting transforms organic waste into a valuable soil amendment, enriching its structure and nutrient content. It's a cornerstone of sustainable agriculture, reducing landfill waste and boosting plant health. Effective composting requires a balance of "browns" (carbon-rich materials like dried leaves and twigs) and "greens" (nitrogen-rich materials such as grass clippings and fruit peels). A proper ratio ensures efficient decomposition, preventing unpleasant odors and maximizing nutrient production.

Several composting methods cater to various needs and space constraints. The simplest is the heap method, involving piling organic matter in a designated area, regularly turning it to aerate the pile. This encourages microbial activity crucial for decomposition. However, this method can be less efficient than others, and requires more space. It might also attract pests if not managed carefully. Regular turning is key to success; a simple pitchfork works well for this task. Aim for a consistently moist, but not soggy, pile.

For smaller spaces or those seeking a more controlled process, bin composting offers a practical solution. Compost bins, ranging from simple wooden structures to more sophisticated models with rotating drums, provide containment and often facilitate better aeration. Bins maintain a higher temperature for faster decomposition, leading to a richer compost in less time. These commercially available bins can also be aesthetically pleasing and easily integrated into a garden setting. Experiment to determine the optimal size and design for your needs, always accounting for your available space.

Vermicomposting, or worm composting, uses red wiggler worms to break down organic matter. This method is particularly suitable for smaller-scale operations. Worms efficiently process waste, creating nutrient-rich castings – essentially worm poop – which is an exceptional fertilizer. While requiring less space than heap composting, it demands specific conditions such as moisture and temperature regulation. Red wiggler worms thrive in a specific environment, requiring careful attention to detail and consistent monitoring. This method is often preferred for its superior nutrient-rich output and compact nature.

Regardless of the chosen method, monitoring the composting process is vital. The pile's temperature should rise during decomposition, indicating active microbial activity. Regular turning ensures even decomposition, preventing anaerobic conditions that produce foul odors and impede the process. Maintain sufficient moisture, but avoid overwatering, and consider adding more browns or greens to adjust the carbon-to-nitrogen ratio as needed. Observation and adjustment are key to successful composting.

Once the composting process is complete, the resulting compost will be dark, crumbly, and earthy-smelling. It should be relatively free from recognizable pieces of the original organic matter. This mature compost is then ready to enrich your garden soil. Compost is best incorporated into the soil before planting, allowing sufficient time for nutrients to become available to plants. Composting isn't just about waste reduction; it's a fundamental step in creating healthy, fertile soil, the foundation of any successful gardening or farming endeavor. Furthermore, understanding these various methods helps you tailor your approach to the specifics of your environment and scale of operation. Remember, patience and attention to detail are key to reaping the rewards of home composting.

2.2 No-Till Farming

No-till farming, a revolutionary approach to agriculture, minimizes soil disturbance. Unlike conventional plowing, which disrupts the soil's delicate structure, no-till methods leave plant residue on the surface, protecting the soil from erosion and the elements. This protective layer acts as a natural mulch, retaining moisture, suppressing weeds, and fostering a thriving ecosystem of beneficial soil organisms. It's a simple concept, yet its impact on soil health, water conservation, and overall farm productivity is profound, offering a sustainable pathway towards a more resilient food system. The benefits extend beyond soil health. No-till farming significantly reduces the need for synthetic fertilizers and pesticides. Healthy soil, teeming with beneficial microbes, naturally provides essential nutrients to plants, reducing our reliance on external inputs. Reduced pesticide use protects beneficial insects and pollinators, promoting biodiversity and creating a healthier environment for both wildlife and humans. Moreover, leaving plant residue on the surface helps to sequester carbon in the soil, mitigating climate change and offering a more environmentally friendly solution to feeding our growing population.

Implementing no-till techniques requires careful planning and adaptation to specific environmental conditions. Choosing the right equipment, like no-till drills and specialized planters, is essential for success. This initial investment may seem daunting, but the long-term benefits far outweigh the initial costs. Farmers can expect reduced fuel consumption, less labor, and potentially higher yields over time as their soil health improves.

However, transitioning to no-till farming isn't a quick fix; it requires patience and a willingness to adapt. Initially, weed management might present a challenge, and careful planning and potentially the use of cover crops are necessary to keep weeds under control. But this initial investment in time and adaptation paves the way for a more sustainable and profitable farming operation.

Furthermore, the transition to no-till often involves learning new techniques and approaches. Farmers need to learn how to manage residue, monitor soil health indicators, and adjust their farming practices accordingly. Accessing workshops, training programs, and engaging with experienced no-till farmers can provide invaluable support and guidance during this transition phase, making the process smoother and more successful. Local agricultural extension services and community networks can serve as excellent resources in this regard.

Beyond the practical aspects, adopting no-till farming promotes environmental stewardship. By reducing soil erosion, improving water retention, and sequestering carbon, this method helps to create a more sustainable and resilient agricultural landscape. It also contributes to a wider movement towards a more environmentally responsible food system, benefiting both the farmer and the planet. The impact extends far beyond the farm itself; it encourages soil health, protects waterways from sediment runoff, and mitigates the impacts of climate change.

Ultimately, embracing no-till farming requires a shift in mindset. It's about moving from a focus on short-term gains to a long-term vision of soil health and sustainable agriculture. It's a commitment to a more holistic approach to farming, one that recognizes the interconnectedness of soil, plants, and the environment. This approach helps farmers build a legacy of fertile land for future generations, ensuring food security and environmental well-being.

It's also important to consider the specific challenges and opportunities that no-till farming presents in diverse contexts. Factors like climate, soil type, and cropping systems all influence the success of no-till practices. For instance, farmers in arid regions may need to adapt their techniques to manage water efficiently, while farmers in areas with heavy rainfall might focus on preventing erosion. Adapting the methods to the environment will enhance its potential and avoid any possible problems.

Therefore, a successful transition to no-till farming involves a commitment to continuous learning and adaptation. Regular soil testing, monitoring crop growth, and experimenting with different techniques are vital for optimizing practices and achieving optimal results. This iterative process allows farmers to refine their approaches, responding effectively to unforeseen circumstances, and continuously refining their farming practices to maximize their sustainability and productivity.

In conclusion, while it necessitates patience and a willingness to adapt, no-till farming represents a promising path towards a more sustainable and resilient agricultural future. By embracing this innovative approach, farmers can contribute significantly to soil health, water conservation, and climate change mitigation, paving the way for a more environmentally friendly and economically viable approach to food production, benefiting both themselves and generations to come. Adopting no-till is a testament to humanity's commitment to food security and environmental sustainability, creating a better future for all.

2.3 Cover Cropping

Cover cropping isn't just about planting something extra; it's a strategic intervention, a clever move in the complex game of soil health. Think of it as a carefully orchestrated performance, with each plant playing a crucial role. Legumes, for example, are the nitrogen-fixing superstars of the cover crop world. Their tiny root nodules house bacteria that transform atmospheric nitrogen into a form plants can readily use, essentially fertilizing the soil for free. This reduces the need for synthetic fertilizers, minimizing environmental impact and lowering costs. Planting a diverse mix of cover crops enhances this benefit, optimizing nitrogen capture.

Rye, with its robust root system, is a master soil builder. Its deep roots penetrate compacted soil, breaking up hardpans and improving drainage. This allows water to infiltrate more easily, reducing runoff and erosion. Furthermore, the extensive root network creates channels for air and water movement, enhancing overall soil structure and aeration. Its dense biomass also acts as a natural mulch, preventing weed growth.

Radishes, meanwhile, are the soil aerators. Their taproots effortlessly punch through compacted layers, creating channels for water and air to flow freely. This improved soil structure increases drainage, allowing for better root penetration of subsequent crops. Consider the benefit to future crops with deeper root systems, accessing nutrients and water more effectively. Their rapid growth makes them ideal for short-term cover cropping.

Beyond these benefits, cover crops offer a robust defense against erosion. Their dense foliage acts as a natural shield, protecting the soil from the erosive forces of wind and rain. This is especially crucial on slopes or in areas with high rainfall. By protecting the soil, cover crops help conserve valuable topsoil. This is essential for long-term soil health and fertility. The reduction in erosion ultimately leads to higher yields and better crop quality.

The timing of planting and termination is crucial for maximizing the benefits of cover crops. Planting in the fall allows them to establish a strong root system before winter, providing significant winter protection against soil erosion and nutrient loss. Spring termination ensures adequate time for decomposition before the next cash crop is planted, supplying nutrients to the soil, in a timely manner, promoting rapid growth and improved yields. Choosing the right mix of cover crops depends on your specific needs and climate. Factors like soil type, climate, and the following main crop should all be considered. Conducting a soil test to determine nutrient levels allows for a targeted cover crop selection. In regions with colder climates, hardy winter rye or cereal rye may be suitable. Warmer areas might benefit from legumes or a diverse mix of species, increasing the range of benefits.

Managing cover crops effectively involves careful planning and execution. Successful cover cropping requires regular monitoring of plant growth and the timely application of any necessary management practices. This may include mowing or rolling to manage biomass, terminating the cover crop at the optimal time, and incorporating it into the soil. This can involve multiple methods depending on the type of cover crop and the tools available.

Remember, successful cover cropping is a journey, not a destination. It involves continuous learning and adaptation. Experimentation with different cover crop species and combinations can optimize your approach over time. Don't be afraid to seek advice from experienced farmers or agricultural extension agents in your area. They offer invaluable insights tailored to your unique conditions.

Cover cropping is more than just a farming technique; it's a holistic approach to soil health and sustainable agriculture. By incorporating cover crops into your farming system, you're investing in the long-term health and productivity of your land. This contributes to environmental sustainability, food security, and economic viability for future generations. Investing in soil health is investing in a sustainable future. The improvement in soil structure and fertility directly translates to increased yields, potentially lowering costs and enhancing profitability. This thoughtful approach to farming makes cover cropping a vital component in creating resilient and sustainable agricultural systems. The rewards extend beyond immediate gains; it's an investment in a more secure and productive future. Ultimately, healthy soil is the foundation of a sustainable food system. And that's a powerful legacy to leave behind.

3 Chapter 3: Water Wisdom

Water is life, especially in agriculture. Efficient irrigation techniques are paramount. Drip irrigation, for instance, delivers water directly to plant roots, minimizing evaporation and runoff, leading to significant water savings compared to traditional flood irrigation. This precision approach not only conserves water but also boosts yields by ensuring each plant receives the exact amount of water it needs. This method is particularly valuable in arid and semi-arid regions, where water resources are scarce. Consider the financial benefits too—reduced water bills translate to increased profitability.

Water harvesting, another crucial strategy, involves collecting and storing rainwater for later use. This can involve simple techniques like constructing swales or building small reservoirs, or more complex systems like rainwater harvesting tanks. These systems provide a reliable water source during dry periods, buffering against drought conditions. Beyond the practical benefits, water harvesting reduces reliance on external water sources, promoting self-sufficiency and resilience in the face of water scarcity. The collected water can then be used for irrigation or other agricultural purposes. The initial investment cost can be offset by the long-term savings on water bills and increased crop yields.

Drought-resistant crops are a game-changer in water-stressed regions. Choosing varieties bred specifically for drought tolerance drastically reduces water consumption while maintaining reasonable yields. Research into drought-resistant crops is constantly evolving, offering a wider selection of options each year. Selecting these varieties requires careful consideration of local climate conditions and soil types. A deeper understanding of your regional climate helps in selecting the best crop varieties for drought conditions. It's a proactive strategy that minimizes vulnerability to water shortages, safeguarding food production, and enhancing long-term sustainability.

Water conservation transcends specific techniques; it's a mindset. Regularly inspecting irrigation systems for leaks, ensuring efficient water use in processing and cleaning, and educating oneself on sustainable water practices are all crucial components. Implementing mulching, a simple practice of covering the soil surface with organic material, can significantly reduce evaporation, keeping the soil moist for longer. These seemingly small steps collectively contribute to significant reductions in overall water usage. A thoughtful and integrated approach to water conservation promotes efficiency and minimizes waste.

Rainwater capture is a powerful tool. It goes beyond simple collection; it's about designing systems that effectively gather, filter, and store rainwater. Rooftop harvesting systems, for instance, can provide a significant source of water for irrigation or household use. Constructing proper drainage systems around fields can also channel rainwater into storage areas, replenishing groundwater resources. Properly maintaining these systems is crucial to ensure effective operation and prevent contamination. This not only conserves valuable water resources, but it also minimizes reliance on municipal water supplies, providing a degree of independence during times of water stress. Moreover, capturing rainwater reduces the strain on municipal water systems, benefiting the community as a whole.

3.1 Efficient Irrigation

Efficient irrigation hinges on understanding your specific needs and context. Before investing in any system, thoroughly assess your soil type, water availability, and the crops you're cultivating. Sandy soils drain rapidly, necessitating more frequent, albeit smaller, water applications compared to clay soils that retain water longer and require less frequent, but deeper, watering. Knowing your soil's water-holding capacity is paramount; it dictates irrigation scheduling and frequency. Different crops have varying water requirements; understanding these individual needs is crucial for maximizing yield and minimizing water waste. A thorough site analysis, encompassing soil testing and climate data, is the cornerstone of effective water management.

Drip irrigation, a localized application method, delivers water directly to the plant's root zone, minimizing evaporation and runoff. This precision targeting enhances water use efficiency significantly, reducing water consumption by as much as 50% compared to traditional flood irrigation. It is particularly beneficial for row crops and orchards, allowing for targeted moisture delivery to individual plants. However, installing a drip irrigation system requires an initial investment in materials and potentially labor. Regular maintenance is also necessary to ensure proper functioning and prevent clogging, necessitating periodic flushing and inspection of the entire system. The long-term savings in water, however, often outweigh the initial costs and maintenance requirements.

Choosing the right irrigation system depends heavily on various factors. For small-scale operations, simple methods like soaker hoses or even hand watering with a watering can might suffice. These are cost-effective and easy to implement, ideal for backyard gardens or small-scale farms. However, for larger areas, more efficient systems such as center pivot irrigation or drip irrigation become more economically viable and practically necessary. Center pivot irrigation, while efficient for large, flat fields, requires significant investment and is not suitable for terrains with uneven topography. The choice is therefore a balancing act between initial investment, operational costs, water efficiency, and the specific scale of your agricultural operations.

Beyond the choice of system, optimizing water use involves implementing water-saving techniques. Mulching helps retain soil moisture, reducing evaporation and the need for frequent watering. Selecting drought-tolerant crops adapted to your region's climate naturally reduces water demand. Crop rotation also plays a crucial role in soil health, improving water infiltration and retention. Regular soil testing informs you of nutrient levels, enabling you to optimize fertilization and avoid overwatering associated with nutrient deficiencies. Simple strategies like adjusting irrigation schedules based on weather forecasts significantly improve water use efficiency and yield, minimizing wasteful watering.

Effective irrigation isn't merely about technology; it's about responsible water stewardship. Regular monitoring of soil moisture levels, using moisture meters or visual cues, helps prevent both under-watering and over-watering. Understanding plant water stress symptoms allows for timely adjustments to irrigation schedules, optimizing water use while maintaining plant health.

Participating in local water conservation programs can provide valuable resources, training, and sometimes even financial incentives to adopt efficient irrigation practices. By embracing integrated approaches combining efficient technologies with responsible water management strategies, you can significantly improve irrigation efficiency, leading to increased yields, reduced costs, and a more sustainable agricultural operation. This holistic approach is crucial for long-term success, ensuring both profitability and environmental responsibility. Proper irrigation isn't just a technique; it's a fundamental aspect of responsible farming, ensuring both sustainable practices and healthy yields.

Implementing efficient irrigation practices is a significant step towards sustainable agriculture. This approach involves selecting the appropriate irrigation method, considering the scale of your operation and your specific conditions. Furthermore, optimizing water usage entails integrating strategies like mulching, choosing drought-tolerant crops, and regularly monitoring soil moisture levels. Finally, adopting water-saving measures helps to reduce water consumption, enhancing both the financial viability and the environmental sustainability of your farming efforts, leading to a healthier planet and increased yields. Remember, the cornerstone of efficient irrigation is understanding your individual needs and consistently adapting your strategies to meet those needs. Ultimately, responsible water management is not just about cost savings, it's about ensuring a secure and productive future for agriculture and for the planet.

3.2 Water Harvesting

Water harvesting, at its core, is about collecting and storing rainwater for later use. This isn't some futuristic technology; ancient civilizations utilized this practice effectively, proving its long-term viability. Think of the ingenious systems developed in arid regions, systems that continue to inspire modern solutions. These techniques aren't simply historical relics; they are practical, low-cost methods with significant potential for boosting agricultural resilience, particularly in regions prone to drought or with limited access to irrigation.

The simplest form of water harvesting involves constructing small-scale systems on your property. Consider building swales, shallow channels dug along contour lines to intercept and slow the flow of rainwater. This allows the water to infiltrate the soil, replenishing groundwater reserves and reducing runoff. A more advanced approach involves creating larger rainwater harvesting systems. This could involve installing rooftop gutters and downspouts to direct rainwater into storage tanks. These tanks can range from simple barrels to larger, purpose-built reservoirs, depending on your needs and available resources. The collected water can then be used for irrigation, reducing your dependence on unsustainable water sources. Remember, careful planning and consideration of local conditions are crucial for efficient water harvesting.

Proper site selection is paramount for success. Identify areas with high rainfall, good soil infiltration rates, and minimal erosion risk. The size of your harvesting system will naturally depend on the size of your land, the intensity of rainfall in your region, and your water requirements. Don't underestimate the importance of proper maintenance. Regular cleaning of gutters, downspouts, and storage tanks is essential to prevent clogging and ensure the longevity of your system. Keeping your storage tanks covered helps prevent contamination and reduces evaporation. In many cases, a simple filter can significantly improve the quality of collected rainwater for irrigation purposes. Thinking ahead and properly maintaining the systems you choose to implement will provide years of fruitful outcomes.

Beyond individual efforts, community-based water harvesting initiatives hold tremendous potential. Collective action can lead to the creation of larger-scale systems, such as community reservoirs or managed aquifer recharge projects. These collaborative projects can provide water security for a wider population, fostering community resilience and economic development. Shared resources and knowledge can significantly improve the sustainability and efficiency of such initiatives. The collaborative aspect of such ventures is crucial; successful implementation relies heavily on community participation and shared responsibility for maintaining the infrastructure. Involving local communities in the design and implementation phases is critical, ensuring the system aligns with local needs and traditional knowledge. For example, working with local communities can help identify and utilize traditional water harvesting techniques that are well-suited to the specific environmental conditions.

Furthermore, water harvesting techniques are not limited to rainwater collection. Consider techniques like fog harvesting, which utilizes specially designed nets to capture moisture from fog, a relevant solution in coastal or mountainous regions. Similarly, dew collection systems can provide a small but steady supply of water in arid climates. These innovative approaches demonstrate the diverse possibilities within water harvesting, offering solutions tailored to specific geographical and environmental conditions. Always consider the specific climate and geological circumstances of your region to choose the method that is most suitable for your environment and your resources. Don't hesitate to collaborate with local experts and agricultural professionals for advice.

Lastly, integrating water harvesting with other sustainable agricultural practices, such as conservation tillage and drought-resistant crop selection, is key to maximizing its benefits. This integrated approach reduces water consumption, improves soil health, and enhances overall agricultural resilience. It forms a holistic strategy for increasing the long-term sustainability of your land and your yield. By adopting a systems-thinking approach, you can develop a truly sustainable water management strategy that supports robust agricultural practices. This is a crucial aspect for sustainable food security in the years to come. Remember, water harvesting is not just a technique; it's a philosophy of responsible water management and a critical component of building resilient food systems, even in challenging environments. This long-term vision should guide all your decisions and practices.

3.3 Drought Tolerance

Developing drought tolerance in crops is paramount for ensuring food security in a world facing increasing aridity. This isn't simply about surviving droughts; it's about thriving despite them, maximizing yields even under water-stressed conditions. We need to move beyond reactive measures and embrace proactive strategies that build resilience into our agricultural systems from the ground up.

One key aspect involves selecting drought-resistant crop varieties. This means carefully studying the genetics of various plants, identifying those with naturally higher tolerance for water scarcity. This isn't just about looking for plants that survive; it's about finding plants that produce a decent yield even under stress. Modern breeding techniques, coupled with traditional knowledge of locally adapted varieties, can play a vital role in developing superior drought-tolerant lines. Farmers can also participate in participatory breeding programs, helping to select the varieties best suited to their specific needs and environments.

Beyond genetics, effective water management is critical. Efficient irrigation techniques, such as drip irrigation or subsurface drip irrigation, can drastically reduce water waste compared to traditional flood irrigation. These methods deliver water directly to the roots, minimizing evaporation and runoff. Precision agriculture technologies, like soil moisture sensors, provide real-time data on soil water content, allowing farmers to optimize irrigation schedules and avoid overwatering or underwatering. These sensors can also help farmers understand their specific soil's water holding capacity. This knowledge will greatly aid in deciding when to irrigate.

Improving soil health significantly boosts drought tolerance. Healthy soil acts like a sponge, retaining more moisture during dry periods. Practices like no-till farming, cover cropping, and composting enhance soil structure and organic matter content. A soil rich in organic matter improves the soil's water retention capacity, leading to reduced water stress on plants. Additionally, cover crops can reduce water evaporation from the soil surface, protecting it from direct sunlight and wind, thereby retaining more moisture. Proper soil management requires regular monitoring of the soil's health status, a vital step for ensuring the soil's capacity to retain water.

Mulching is a simple yet powerful technique to improve drought tolerance.

A layer of mulch around plants shades the soil, reducing evaporation and maintaining soil moisture. Organic mulches, such as straw or wood chips, also improve soil health over time, providing further benefits. Selecting the right type of mulch depends on various factors such as the climate and the type of plants being grown. Farmers should carefully consider the characteristics of various mulching materials to decide what works best for their conditions. This knowledge requires a balance between theoretical and practical understanding.

Drought-resistant cultivars aren't a silver bullet. Their success hinges on adopting integrated approaches that consider every aspect of the farming system. Implementing these methods requires a shift in mindset, from focusing solely on yield maximization to prioritizing resilience and sustainability. A holistic approach involves careful consideration of all these factors, which is why a combined effort is needed for optimal results.

Beyond these practices, there's a need for ongoing research and development in drought tolerance. This involves exploring the potential of biotechnology and genetic engineering to further enhance crop resilience. Scientists are continually working to improve plant breeding technologies to develop cultivars that are capable of withstanding more extreme conditions. This research also involves exploring the potential of other strategies to enhance the drought resilience of crops, such as the development of drought-tolerant root systems. This work is vital for developing sustainable agricultural practices.

Furthermore, community-level initiatives are crucial. Sharing knowledge and best practices among farmers is essential for scaling up effective drought management strategies. This might involve farmer field schools, workshops, or online forums, where farmers can share their experiences and learn from each other. These are not just about theoretical knowledge but also about the practical experiences of successful farmers. The exchange of these experiences is extremely important in adopting new strategies.

Finally, effective drought management requires proactive planning and preparedness. This includes regular monitoring of weather patterns and soil moisture levels, as well as the development of contingency plans for dealing with droughts. Early warning systems and access to drought insurance can greatly assist farmers in mitigating the effects of dry periods. This forward-looking approach allows for better resource allocation, minimizing the impacts of drought events. The availability of resources and support plays a crucial role in ensuring farmers are capable of effectively handling water scarcity situations. Effective drought preparedness measures, therefore, are crucial for agricultural sustainability.

3.4 Water Conservation

Water conservation is paramount for sustainable agriculture. Think of it this way: every drop counts. Efficient irrigation techniques, like drip irrigation or micro-sprinklers, deliver water directly to plant roots, minimizing evaporation and runoff—a stark contrast to wasteful flood irrigation. Implementing these methods can significantly reduce water consumption without sacrificing yield. Consider the long-term benefits: lower water bills, reduced environmental impact, and enhanced resilience against drought. It's an investment in your farm's future.

Beyond irrigation, water harvesting techniques offer another avenue for conservation. This involves collecting rainwater from rooftops, paved surfaces, and even gently sloping land. Simple structures like rainwater tanks or strategically placed swales can capture and store significant amounts of water, providing a supplementary source for irrigation during dry spells. These systems not only conserve water but also reduce reliance on external water sources. Furthermore, water harvesting can help recharge groundwater levels, a vital resource for long-term sustainability. Planning and installation are straightforward, making it accessible to many farmers.

Selecting drought-tolerant crops is another powerful strategy. Many crops, like certain varieties of sorghum, millet, and beans, possess inherent drought resistance. These crops require less water to thrive, making them ideal for arid or semi-arid regions. Researching and selecting appropriate varieties tailored to your specific climate and soil conditions is crucial. This approach reduces water stress on crops, ensuring better yields even with limited rainfall. Additionally, focusing on these resilient plants reduces the overall water footprint of your farm, making it more sustainable.

Rainwater capture is a relatively simple, yet highly effective, water conservation method. It's about collecting rainwater runoff from your fields, directing it into storage for later use. This captured water, though perhaps initially appearing unsuitable, can be utilized effectively with minimal treatment. By integrating this into your farming practice, you're not only preserving a vital resource, but also actively replenishing the soil moisture levels, which indirectly improves the health of the land and its yields. It demonstrates a proactive approach to water management that complements other techniques.

Effective water conservation isn't merely about implementing individual techniques; it's about a holistic approach. Regular monitoring of soil moisture levels using tools like soil moisture sensors helps in optimizing irrigation schedules, preventing overwatering. This data-driven approach leads to more efficient water use and prevents wastage. Combining this with practices like mulching, which helps retain soil moisture, significantly improves water conservation strategies. Moreover, educating oneself on the latest research and best practices, and sharing knowledge with fellow farmers, ensures the most effective and sustainable water use. It's about the entire community learning and growing together.

3.5 Rainwater Capture

Rainwater harvesting offers a practical, cost-effective solution to water scarcity, a critical challenge for sustainable agriculture. Implementing a rainwater harvesting system involves several key steps, beginning with a thorough assessment of your needs. Consider the size of your garden, the average annual rainfall in your region, and the water requirements of your crops. This careful planning will ensure your system is adequately sized and efficient.

Understanding your roof's catchment area is vital. Measure the dimensions of your roof to calculate its surface area; this will determine the potential volume of water you can collect. Remember to account for factors that might reduce collection, such as trees shading the roof or debris accumulation. A well-maintained roof is crucial for optimal water quality.

Choosing the right collection system depends on several factors. Larger systems might involve constructing gutters, downspouts, and storage tanks, possibly underground. For smaller gardens, rain barrels are a popular and accessible option. These barrels are relatively inexpensive and readily available from many garden supply stores. Regardless of the size, ensure your chosen system is durable and resistant to leakage. Regular maintenance will prolong its lifespan.

Once you've collected the rainwater, you need a way to effectively use it. This could involve a simple gravity-fed system where the water flows directly to your plants, or a more complex system with pumps and irrigation lines for precise control. Consider the type of plants you grow, as different plants have varying water requirements. Overwatering can be as damaging as underwatering.

Water quality is a significant factor. Rainwater, while naturally soft, can still contain pollutants. Installing a filter can remove debris and improve water quality, protecting your plants from contaminants. Simple filtration, such as screening, can be adequate in many cases. Consider the levels of pollutants common in your area and opt for a filter accordingly.

The storage aspect is equally important. Choose a tank or barrel material appropriate for your environment and budget. Materials such as food-grade plastic or stainless steel are commonly used, offering durability and preventing contamination. Proper sizing of your storage is essential, providing sufficient capacity to meet your needs throughout drier periods. Regular cleaning of the storage container prevents the build-up of algae or bacteria.

Incorporating rainwater harvesting into your overall water management strategy is essential. This might involve a combination of rainwater and other water sources like well water or municipal supply. This integrated approach ensures resilience, particularly during periods of low rainfall. Regular monitoring of your system's performance allows for adjustments to maximize efficiency.

Beyond the practical aspects, remember the environmental benefits. Rainwater harvesting reduces reliance on municipal water supplies, conserving this precious resource for other uses. It can also lessen the burden on local water systems, particularly in regions experiencing water stress. By reducing water consumption, you lessen the environmental impact of your gardening practices.

Finally, consider the potential challenges. Initial costs can be a barrier for some, although government subsidies or rebates might be available in certain regions. Maintenance is also required, ensuring the system's continued effectiveness. Despite these challenges, the long-term benefits of rainwater harvesting often outweigh the initial investment and effort. It is an investment in sustainable agriculture and water security. The rewards are many – healthier plants, reduced water bills and a reduced environmental footprint; all valuable assets in a world facing growing water scarcity. Careful planning, appropriate system selection, and regular maintenance are crucial to the success of your rainwater harvesting project, reaping the rewards of a more sustainable gardening approach. The transition may seem daunting, but the advantages are well worth the initial commitment and effort involved. With a little research and planning, you can effectively implement a rainwater harvesting system and contribute to a more resilient and sustainable future for your gardening efforts.

4 Chapter 4: Community Gardens

Community gardens offer a powerful model for fostering local food systems and strengthening community bonds. They're more than just plots of land; they're vibrant hubs of activity, transforming neglected spaces into thriving ecosystems. Imagine a once-barren lot now bursting with life, a testament to collective effort and shared harvests. This isn't just about growing food; it's about nurturing a sense of belonging and shared responsibility.

The practical benefits are immediately apparent. Increased access to fresh, healthy produce is a significant advantage, especially in food deserts where supermarkets are scarce. Residents gain a direct connection to their food source, understanding the journey from seed to table in a way that's rarely possible with commercially grown produce. This firsthand experience fosters appreciation for the effort involved in food production and promotes healthier eating habits.

Furthermore, community gardens can revitalize neighborhoods. They improve aesthetics, enhancing property values and creating attractive public spaces. The collaborative nature of these gardens promotes social interaction, bringing neighbors together in a shared purpose. This fosters a sense of community, building trust and strengthening social networks—a crucial factor in building resilient and supportive environments.

Beyond the social aspects, community gardens contribute significantly to environmental sustainability. They often incorporate ecological practices like composting and water conservation, reducing reliance on harmful chemicals and minimizing environmental impact. These gardens can also serve as micro-habitats, supporting local biodiversity and providing green spaces within urban environments. The cumulative effect of many such gardens across a city can be substantial, contributing to a greener, more sustainable urban landscape.

Starting a community garden requires careful planning and community engagement. Finding a suitable location, securing necessary permissions, and obtaining resources are initial steps. This often involves navigating local regulations and collaborating with city officials or landowners. Successful projects prioritize inclusive participation, ensuring the garden reflects the diversity of the community it serves. Workshops and training sessions can empower participants with the necessary gardening skills, from seed starting to pest management.

Securing funding is another key consideration. Grants, sponsorships, and fundraising events can provide crucial financial support. Innovative approaches, such as community-supported agriculture (CSA) models, can generate income and ensure the garden's long-term viability. Establishing clear governance structures and volunteer roles is vital for effective management and sustainability. Regular maintenance, conflict resolution, and consistent communication are essential for long-term success.

The long-term impact of community gardens extends beyond the immediate benefits. They serve as educational platforms, teaching valuable skills and fostering environmental awareness. Children participating in the garden develop a deeper connection with nature, learning about the food cycle and the importance of sustainable practices. These gardens also promote intergenerational learning, connecting younger generations with the wisdom and experience of older community members.

Community gardens aren't simply a trend; they represent a fundamental shift towards more sustainable and equitable food systems. They offer a tangible way for communities to take control of their food security, improve their health, and strengthen their social fabric. By embracing the principles of collaboration, sustainability, and community engagement, these gardens have the potential to transform urban landscapes and nourish both bodies and spirits. The success stories emerging from across the globe offer compelling evidence of their power to create positive change, demonstrating the potential for community gardens to be a vital component of a more just and sustainable future. Their creation requires commitment and collaboration, but the rewards – both tangible and intangible – are immeasurable. The vibrant tapestry of life that unfolds within a community garden is a powerful symbol of the human spirit's capacity for resilience, innovation, and collective action. Their enduring legacy lies not just in the food they produce, but in the strengthened community they cultivate, demonstrating that sustainable food systems and vibrant communities are inherently linked, each reinforcing and enriching the other.

4.1 Urban Farming

Rooftop farms, even small ones, can significantly boost local food production. Imagine a vibrant patch of tomatoes basking in the sun, their juicy fruit ripening just above the city's clamor. This isn't a fantasy; it's a realistic and achievable goal for many urban dwellers. With careful planning and a little effort, you can transform unused spaces into productive gardens, providing fresh, healthy food while contributing to a more sustainable urban environment. Consider the space you have available – a balcony, a rooftop, even a window box – and research the best crops for your chosen location and the amount of sunlight it receives.

Vertical gardening is a game-changer for urban environments where space is at a premium. Think of walls adorned with climbing beans or strawberries cascading from hanging baskets. These systems maximize space utilization, allowing you to grow a significant amount of food in a relatively small area. There are numerous vertical gardening systems available, from simple DIY designs to sophisticated hydroponic setups. Research what best fits your budget, skills, and the specifics of your space. Don't be afraid to experiment and adapt your design as you gain experience. Remember that consistent monitoring is key to success.

Community gardens represent a powerful force in fostering urban food production and community engagement. These collaborative spaces offer a platform for people to connect, share knowledge, and work together to cultivate food. Joining a local community garden can provide valuable learning opportunities, access to shared resources, and the satisfaction of contributing to a larger, shared goal. If a nearby community garden doesn't exist, consider initiating one yourself. Gather your neighbors, identify a suitable location, and work together to transform a vacant lot or underutilized space into a thriving hub of community and food production. It may require navigating local regulations and securing permissions but the rewards are far-reaching.

Composting is essential for urban farming. This process not only reduces waste but also creates nutrient-rich soil amendments crucial for healthy plant growth. In apartment settings, a small indoor worm composting bin is effective. For those with access to outdoor space, a traditional compost bin offers a larger capacity. Regardless of your chosen method, make sure you understand the basic principles of composting, including layering green and brown materials and maintaining appropriate moisture levels. Remember consistent aeration is important for the decomposition process. Proper composting helps eliminate food waste from landfills and drastically reduces your reliance on commercially produced fertilizers.

Choosing the right plants is critical for urban farming success. Opt for varieties known for their resilience and adaptability to diverse conditions. Many heirloom varieties, often passed down through generations, possess excellent disease resistance and thrive even with limited resources. They're often more flavorful than commercially produced plants as well. Research varieties well-suited for containers and vertical gardening, and select plants that can withstand the specific environmental conditions in your urban setting, whether that's limited sunlight or exposure to wind. Remember to consider companion planting, a technique that maximizes growth and health by planting specific crops together.

Urban farming presents challenges, but the rewards far outweigh the effort. Overcoming these obstacles involves learning about pest and disease control specific to your region and the types of plants you grow. Natural pest control methods, such as companion planting and introducing beneficial insects, are environmentally friendly and effective alternatives to chemical pesticides. Regular plant inspection can help identify issues early on, allowing for swift action before problems become widespread. Furthermore, knowledge of local microclimates can help you choose the most successful planting locations for your specific crops.

Harvesting your urban crops is a rewarding experience, symbolizing the fruits of your labor and the connection to your food. Preservation methods are crucial for maximizing the yield and enjoying your harvest throughout the year. Techniques such as freezing, canning, and drying extend the life of your produce, providing a continuous supply of fresh, homegrown food. Learning these techniques not only enhances your farming experience but also helps reduce food waste and ensures you have access to nutritious food year-round. Experiment with different methods, and consider what will best suit your needs.

Beyond the practical aspects, urban farming offers immense psychological benefits. Connecting with nature, nurturing plants, and witnessing the growth cycle can foster a sense of calm and well-being. This connection transcends the simple act of food production; it's a path towards a more meaningful and sustainable lifestyle. Moreover, the process offers opportunities for learning, self-sufficiency and personal growth. It is rewarding, educational, and contributes positively to the environment.

The future of urban farming is bright, fueled by innovation and increasing awareness of its numerous benefits. Technological advancements, such as hydroponics and aeroponics, offer new possibilities for maximizing food production in limited spaces. As urban populations continue to grow, the role of urban farming in ensuring food security and promoting sustainable living will only become more significant. This trend provides an exciting opportunity to reshape our relationship with food, fostering a more resilient and interconnected urban environment. Remember, continuous learning and adaptation are keys to thriving within the changing landscape of urban farming.

Urban farming, therefore, is more than just growing food; it's about building community, fostering sustainability, and reclaiming a connection with nature within the urban environment. It's a journey of learning, adaptation, and abundant rewards. From rooftop gardens to community plots, urban farming empowers individuals and communities alike to actively participate in creating a more resilient, healthy, and sustainable future. Start small, learn consistently, and witness the transformative power of urban farming.

4.2 Shared Resources

Sharing resources within a community garden isn't just about pooling seeds or tools; it's about cultivating a shared sense of responsibility and mutual support. Imagine a network, where experienced gardeners mentor newcomers, passing down invaluable knowledge—from the optimal planting time for specific crops to effective pest control methods using natural remedies. This collaborative spirit fosters a sense of belonging, transforming the garden into a vibrant hub of shared learning and accomplishment. It's a powerful antidote to the isolation that can often accompany modern life, replacing it with a feeling of connection and common purpose. The garden becomes more than a place to grow food; it becomes a vital social fabric weaving together the threads of a community.

Effective resource sharing requires clear communication and well-defined systems. A shared calendar, for instance, can prevent scheduling conflicts over the use of tools or plots. Designated storage areas for communal supplies like compost, tools, and fertilizers ensure everything is easily accessible and well-maintained. Implementing a simple inventory system—even a shared spreadsheet—keeps everyone informed about what's available and what needs replenishing. These seemingly minor details are crucial for seamless operation, promoting both efficiency and goodwill. The more transparent and well-organized the system, the smoother the collaboration will be, preventing potential friction and misunderstandings among participants.

Beyond the tangible resources, successful community gardens share knowledge. Regular workshops or informal gatherings can serve as platforms for exchanging gardening tips, problem-solving, and sharing recipes. These events foster a stronger sense of community and provide opportunities for mutual learning. A seasoned gardener might demonstrate a unique composting technique; a newcomer might introduce a new variety of heirloom tomato. This constant exchange of ideas and experiences enriches the garden and strengthens the bonds between its members. The result is a collective wisdom that surpasses any individual's knowledge, a dynamic resource far more valuable than any single tool or seed.

However, successful resource sharing requires careful consideration of equity and access. Ensuring everyone has equal opportunity to participate and benefit from the shared resources is paramount. This might involve creating designated plots for individuals with limited mobility or designing tasks that accommodate diverse abilities. Equitable access to the garden's bounty—perhaps through a system of shared harvests or a community potluck—is equally crucial, preventing any individual or group from disproportionately benefiting from the collective effort. The goal is to foster an inclusive environment where everyone feels valued and empowered. Only then can the true potential of shared resources be fully realized.

Financial management is another critical aspect of shared resource success. Transparency is key. A clear budget, regularly reviewed and discussed among members, ensures everyone understands how funds are allocated. This might involve setting membership fees, securing grants, or organizing fundraising events. Open communication regarding finances builds trust and prevents misunderstandings, promoting both financial stability and community cohesion. A well-managed financial system contributes significantly to the long-term sustainability and success of the community garden, eliminating a major source of potential conflict and ensuring that the garden flourishes for years to come.

Finally, establishing a clear decision-making process is essential for the long-term health of the community garden. This could involve regular meetings, where members discuss issues and vote on proposals. Agreeing on a set of bylaws—a document outlining the garden's rules and responsibilities—provides a framework for resolving disagreements fairly. Establishing a clear leadership structure can streamline decision-making, avoiding potential conflicts and ensuring efficient management. A robust decision-making process is the backbone of a successful community garden, ensuring that everyone's voice is heard and that the garden thrives as a truly collaborative endeavor. Building a strong foundation of shared governance is as important as sharing seeds and tools.

5 Chapter 5: Seed Saving

Saving seeds isn't just about preserving a few extra for next year's planting; it's about safeguarding biodiversity, securing your food future, and reclaiming control over your food supply. Understanding the nuances of seed saving is crucial for building resilient food systems. It's a skill passed down through generations, a testament to humanity's enduring connection to the land, and it's a skill everyone can learn.

Seed selection is paramount. You wouldn't choose a sickly plant to propagate, would you? Similarly, choose seeds from robust, healthy plants showing exceptional traits – high yield, disease resistance, or desirable flavor profiles. Look for plants that have successfully adapted to your specific local conditions, ensuring the best chance of success for future generations. Avoid seeds from plants displaying any signs of disease or weakness; you want strong, resilient genetics in your next harvest. Document your choices; noting the plant's characteristics is essential for future reference, helping maintain the quality of your seed bank over time. It's about building a legacy of strong plants.

Proper seed storage significantly impacts longevity. Seeds are living organisms, and their viability diminishes over time, influenced by moisture, temperature, and oxygen exposure. Low temperatures and low humidity are key—aim for a cool, dry location ideally around 40°F (4°C) with relative humidity below 30%. Airtight containers—glass jars are excellent—protect seeds from moisture and pests. Label each container clearly with the seed type and harvest date. Regularly check for any signs of pest infestation or deterioration, promptly removing any compromised seeds to prevent spreading. A well-maintained seed bank is an invaluable resource—a life raft in times of need.

Open-pollinated seeds are the cornerstone of seed saving. These seeds, unlike their hybrid counterparts, faithfully reproduce the parent plant's characteristics when grown. This means saving seeds from these plants will yield offspring true to type, generation after generation. This characteristic is crucial for maintaining genetic diversity and adapting crops to specific local environments. Preserving these seeds is essential for safeguarding the integrity of our food supply and promoting agricultural resilience. This is not simply a matter of convenience; it's a critical step toward long-term food security.

Hybrid seeds, on the other hand, are created by crossbreeding two different parent plants. While they often exhibit superior traits like high yield and disease resistance in the first generation (F1), subsequent generations (F2 and beyond) show increased variability, often losing the desired hybrid characteristics. Saving seeds from hybrid plants will not yield the same quality, reliability, or uniformity as the initial generation, limiting their practicality for long-term seed saving. For this reason, seed savers mostly focus on open-pollinated varieties.

Maintaining seed purity is essential for preserving the desirable characteristics of your chosen varieties. This means carefully isolating your plants during flowering, preventing unintended cross-pollination. Spatial separation is vital, particularly with plants that share pollination vectors, like wind or insects. Physical barriers, like netting or row covers, can also be effective. Closely monitoring your plants during flowering is essential to detect any unwanted pollination events. This conscientious approach is vital for maintaining the integrity of your seed bank and ensuring the long-term success of your seed saving efforts. Failing to maintain purity leads to a degradation of traits and a loss of the quality of seed stock.

Seed saving isn't just a practice; it's an act of resilience, a powerful tool for building food sovereignty and ensuring a sustainable future. It connects us to our agricultural heritage, promotes biodiversity, and empowers individuals to become active participants in shaping their food systems. By actively engaging in seed saving, you're not just planting seeds; you're planting the seeds of a more resilient, equitable, and secure future for your community and beyond. It is a legacy you create, a living history that provides for generations to come. It's a commitment to a future where food independence is within everyone's reach. The knowledge you gain, the seeds you save, these are not mere actions; they are investments in a sustainable future.

5.1 Seed Selection

Choosing the right seeds is paramount; it's the foundation upon which your entire harvest rests. Don't underestimate this seemingly simple step. It's not just about picking a pretty packet; it's about understanding your climate, your soil, and your goals. Selecting seeds that thrive in your specific environment guarantees a higher yield and less frustration. Ignoring this crucial aspect can lead to wasted time, resources, and ultimately, a disappointing harvest.

Consider your climate's nuances. Are you in a region with short, cool summers, or do you enjoy long, hot growing seasons? Some seeds, like certain varieties of beans, are far better suited for warmer areas than those requiring shorter growing cycles, such as many lettuce varieties. Understanding your local frost dates is crucial. Planting frost-sensitive crops before the last expected frost will result in losses, and likewise, planting cold-hardy crops too late might mean they won't reach maturity before the first frost. Soil type significantly impacts seed selection. Sandy soil drains quickly, demanding moisture-retaining seeds. Conversely, clay soil, often dense and poorly draining, benefits from seeds that tolerate moist conditions, though well-drained soil is ideal for most seeds. Testing your soil's pH is also essential. Different plants thrive in varying pH levels, so ensure your chosen seeds match your soil's acidity or alkalinity. A simple soil test kit readily available from most garden centers will provide this vital information.

Your gardening goals also play a critical role. Are you aiming for a bountiful harvest for preservation, a continuous supply of fresh vegetables throughout the growing season, or simply a visually appealing garden? This determines the type of seeds you'll prioritize. For instance, heirloom varieties, often favored for their unique flavors and adaptability, might not yield as much as modern hybrids bred for maximum productivity. Choosing between heirloom varieties and hybrid seeds should be based on a thorough understanding of your personal priorities and what you hope to achieve.

Think about the scale of your operation. Starting small with a few seeds of different varieties allows experimentation and helps you learn what thrives in your specific environment. As you gain experience, you can increase the variety and quantity of your seeds, expanding your garden's output accordingly. Don't get overwhelmed by trying to plant everything at once; start small, learn, and build your skills and your garden gradually.

Beyond climate, soil, and goals, consider seed sourcing. Reputable seed companies often provide detailed information about each seed's characteristics, including germination rates and growing requirements. Locally sourced seeds, ideally from a nearby seed saver or farmer, are frequently better adapted to your region's specific conditions. When possible, choose open-pollinated seeds; these retain their genetic diversity, making them crucial for maintaining biodiversity and safeguarding future harvests.

Purchasing seeds requires diligence. Check the packaging for crucial information: the germination rate (a percentage indicating how many seeds are expected to sprout), the planting depth, and the spacing requirements. These details significantly influence the success of your planting efforts. A low germination rate may suggest poor seed quality, prompting you to seek a different source. Don't hesitate to ask questions; reputable suppliers will gladly guide you through seed selection.

Finally, remember that seed selection is an iterative process. Experiment, learn from your successes and failures, and adjust your selection accordingly. Every year offers a new opportunity to refine your approach, tailoring your seed choices to achieve optimal results in your unique growing environment. Keep detailed records of what you plant, when, and the results you achieve. This valuable information will inform your seed selections in subsequent years, leading to increasingly successful harvests.

5.2 Seed Storage

Proper seed storage is crucial for maintaining seed viability and ensuring successful germination. Think of it as preserving a piece of the future harvest, protecting the genetic potential of your crops against the ravages of time and improper conditions. Failure to store seeds correctly often leads to decreased germination rates, a frustrating and costly outcome for any grower. Investing time in understanding and implementing these techniques safeguards your efforts and ensures the success of future plantings.

Seed longevity depends on several factors, most importantly moisture content. High moisture levels encourage mold growth and fungal infections, leading to seed decay and loss of viability. Low moisture content is key. The ideal moisture content for most seeds is between 5% and 7%, significantly reducing the risks associated with increased humidity. This precise control is the bedrock of successful long-term seed storage. Achieving this requires careful drying and sealing techniques, which will be discussed in detail later.

Before storage, properly dry seeds thoroughly. This involves spreading them thinly on a clean, well-ventilated surface, away from direct sunlight and excessive heat. A dehydrator or properly ventilated area in a warm, dry climate can accelerate the drying process. Regularly turning the seeds ensures even drying. Remember, even slightly moist seeds are significantly more vulnerable to spoilage, so thorough drying is paramount. Under-dried seeds can lead to premature germination or rot even in seemingly perfect storage conditions, undermining your efforts.

Once adequately dried, select appropriate containers for storage. Glass jars with airtight lids are an excellent choice; they provide good protection against moisture and pests. Avoid using plastic containers, as they can leach chemicals into seeds over time. Metal containers are acceptable, provided they are rust-proof and well-sealed. The choice of container directly impacts seed preservation, as it determines their vulnerability to environmental factors and pests. Investing in high-quality, air-tight containers is a wise long-term investment.

The storage environment significantly influences seed viability. Keep seeds in a cool, dark, and dry place. Ideal temperature ranges typically fall between 32°F (0°C) and 41°F (5°C). Fluctuations in temperature and humidity should be minimized, as these variations can accelerate seed aging. Consider using a refrigerator or a cool, dry basement for optimal storage conditions. Monitoring the temperature and humidity levels can also help you identify and adjust storage locations as needed.

Regular inspection of your stored seeds is vital. Periodically check for signs of mold, pest infestation, or any other deterioration. If you notice any problems, promptly remove the affected seeds to prevent contamination. A proactive approach minimizes losses and protects the remaining seeds. It's also advisable to maintain detailed records of each seed variety, including the date of harvest, drying method, and storage conditions. This documentation facilitates efficient tracking and allows for adjustments and improvement of your techniques.

Labeling your seed containers is crucial for organizational purposes. Clearly label each container with the seed variety, the date of harvest, and any relevant information, such as treatment. This ensures that you can easily identify and retrieve specific seeds when the time comes to plant. Careful labeling avoids confusion and streamlines the planting process.

In addition to the above, consider using seed packets designed for long-term storage. These packets often feature moisture-absorbing materials and are sealed to create an environment that is inhospitable to pests and moisture. Though they may appear to be a small detail, these packets represent a significant improvement over using improvised containers for seed storage, particularly given the increased viability offered by utilizing appropriate packaging. This seemingly small detail can extend the lifespan of your seeds dramatically.

Proper seed storage practices are not mere technicalities. They represent a fundamental investment in the future of food production, safeguarding biodiversity and ensuring agricultural resilience. By consistently applying the techniques outlined above, you can preserve your seed collection, ensuring a reliable supply of seeds for future plantings. Remember, proper seed storage is more than just preserving seeds; it's about preserving your ability to grow your own food, a fundamental act of self-sufficiency. The preservation of seeds is not merely an agricultural technique; it's a fundamental act of resilience, ensuring the continuance of your food supply for generations to come. Your work ensures the successful propagation of crops, maintaining genetic diversity, and securing future harvests. This crucial process ensures the ongoing health and viability of your crops, reinforcing your efforts towards sustainable and self-sufficient food production.

5.3 Open Pollinated Seeds

Open-pollinated seeds are the cornerstone of biodiversity and agricultural resilience. Unlike their hybrid counterparts, these seeds, when harvested and replanted, produce offspring that are largely true to the parent plant's characteristics. This predictable inheritance allows farmers to maintain desirable traits within their crops year after year, fostering adaptation to local climates and conditions. This is a critical factor in sustaining diverse agricultural landscapes and preserving genetic heritage. It's not just about maintaining the status quo, but actively participating in the evolution of your crops. Consider this a partnership built on mutual respect and understanding.

The process of saving open-pollinated seeds is deceptively simple yet profoundly rewarding. It involves carefully selecting the healthiest and most productive plants from your harvest. These should exhibit the traits you value most – disease resistance, yield, taste, and appropriate maturity for your climate. Remember, you're actively shaping the future of your garden, one seed at a time. This conscious selection process is a direct contribution to the genetic vitality of your crops, safeguarding against potential threats. You are, quite literally, engineering your food security.

Proper seed storage is paramount for preserving seed viability and maintaining genetic purity. Keeping seeds dry, cool, and dark is essential. Airtight containers in a cool, dark, and dry place are ideal. The goal is to inhibit germination and prevent the growth of fungi or other microorganisms that can damage or destroy the seeds. Think of your seed storage as a tiny time capsule, protecting the legacy of your crops for future generations. The simple act of careful storage is a testament to your commitment to long-term sustainability. Neglecting this crucial step would render the rest of the efforts useless.

Maintaining seed purity is crucial in preserving the traits of open-pollinated seeds. Isolation is paramount; preventing cross-pollination with other varieties of the same species is key. This means maintaining a sufficient distance between different varieties to avoid accidental mixing of pollen. Knowing the pollination habits of your crops is essential; some plants are self-pollinating, while others rely on insects or wind. The dedicated isolation of seed varieties prevents the dilution of valuable genetic traits. This might involve spatial separation, specific planting times, or hand-pollination in some cases. The added effort is well worth the preservation of unique characteristics.

The benefits of using open-pollinated seeds extend far beyond the garden. These seeds empower farmers and gardeners to become more independent and resilient in the face of external pressures. They reduce reliance on commercial seed companies, fostering self-sufficiency and reducing environmental impact associated with large-scale seed production and distribution. Choosing these seeds is a deliberate act, a small step in re-connecting with the land and your food source, a conscious effort towards reclaiming control and understanding. This self-reliance builds community, strengthens local economies, and enhances the robustness of food systems. It's about more than just the seeds; it's about food sovereignty and community empowerment.

Open-pollinated seeds are not just seeds; they are a legacy. They represent a connection to the past, a commitment to the present, and a hope for the future. Saving and sharing them becomes an act of stewardship, a tangible contribution to preserving biodiversity and fostering agricultural resilience in an increasingly uncertain world. They are a reminder of our connection to the earth and our responsibility to nurture it. They are a symbol of hope, a testament to the enduring power of nature, and a celebration of human ingenuity. They are a practical, tangible step towards ensuring food security for generations to come, a small act with profound implications. Preserving these seeds is an investment in a sustainable future. It's an act of hope, a testament to the enduring power of nature, and a celebration of human ingenuity. It's a legacy we can all share in and be proud of. The future of our food supply might well depend on it.

5.4 Hybrid Seeds

Hybrid seeds offer significant advantages, particularly increased yield and uniformity. Farmers often choose them for their predictable performance, resulting in crops that mature at the same rate and possess similar characteristics, simplifying harvesting and processing. This consistency is highly valued in commercial agriculture, where efficiency is paramount. However, understanding their limitations is crucial for responsible agricultural practices.

The primary drawback of hybrid seeds is their inability to reproduce true to type. Unlike open-pollinated seeds, the offspring of hybrid plants won't necessarily inherit the desirable traits of their parents. This means farmers must purchase new seeds each year, creating an economic dependence on seed companies. This reliance can limit farmers' autonomy and potentially increase their production costs over time. Consider the long-term implications before committing to solely relying on hybrids.

For smaller-scale farmers or those prioritizing seed sovereignty, the cost and dependence associated with hybrid seeds might outweigh the benefits. The increased yields might not always justify the recurring expense, especially when considering the potential for developing resilient, open-pollinated varieties suited to their specific environment. Many farmers find that a balance, incorporating both hybrid and open-pollinated seeds into their strategy, allows for flexibility and resilience.

Careful consideration of your farming goals is essential. If maximum yield and uniformity are the primary objectives, and financial resources are readily available, hybrid seeds might be a suitable choice. But if budget constraints, environmental concerns, or a desire for long-term seed independence are important factors, a more diversified approach incorporating open-pollinated varieties should be considered.

Understanding the genetic makeup of hybrid seeds is key. Hybrid vigor, or heterosis, results from crossing two distinct inbred lines. This process often leads to plants with superior characteristics compared to their parents. However, this advantage is lost in subsequent generations due to the segregation of genes. This inherent characteristic underscores the need for annual seed purchases. This isn't necessarily a negative, as it can encourage innovation and improvement in seed varieties over time.

Many factors influence the choice between hybrid and open-pollinated seeds. Climate conditions, soil type, pest prevalence, and market demands all play a role in making an informed decision. It's beneficial to conduct thorough research on the specific hybrid varieties suitable for your region and farming practices. Consult with local agricultural experts to acquire tailored advice considering your specific circumstances. Remember, local knowledge and experimentation are invaluable assets.

Furthermore, the environmental impact of relying heavily on hybrid seeds requires careful consideration. The increased use of pesticides and fertilizers often associated with hybrid agriculture can negatively impact biodiversity and soil health. This can affect the long-term sustainability of farming practices. A balanced approach, integrating diverse cropping systems and sustainable farming techniques, helps mitigate these potential drawbacks. There's no one-size-fits-all solution, stressing the importance of making an informed decision based on your situation.

Choosing between hybrid and open-pollinated seeds isn't just about yield. It's about the long-term health of the soil, the resilience of your farm to changing conditions, and your overall approach to sustainability. Hybrid seeds represent a powerful tool in modern agriculture, but their effective use depends on a deep understanding of their strengths and limitations. Understanding this nuance ensures that you can make a choice that aligns with your overall farming philosophy and long-term goals. Making a well-informed decision guarantees you can contribute to a healthier, more sustainable agricultural landscape. Remember, responsible seed selection is a cornerstone of resilient food systems.

5.5 Seed Purity

Maintaining seed purity is paramount for preserving desirable traits and preventing genetic contamination. This is especially crucial when saving seeds from open-pollinated varieties, as cross-pollination can easily occur, leading to a gradual shift in the plant's characteristics over successive generations. Imagine your prized heirloom tomato, slowly losing its unique flavor and becoming indistinguishable from a common supermarket variety due to unwanted pollen drifting on the wind. Protecting this genetic integrity requires careful planning and execution, far beyond simply collecting the seeds.

The isolation distance is a key factor in maintaining purity. Different plant species have varying pollen dispersal ranges; some plants self-pollinate effectively, while others rely on wind or insects for pollination, potentially leading to accidental cross-pollination with neighboring plants. Understanding these pollination mechanisms is vital for determining appropriate isolation distances. For instance, a wind-pollinated crop like corn may necessitate greater isolation than a self-pollinating plant like beans. Researching the specific requirements of your chosen plant species is crucial before beginning the seed-saving process, lest you undermine your efforts before they even truly start. Failing to plan for adequate space and isolation can compromise even the most meticulously collected seeds.

Careful hand-pollination can mitigate some risks. This method involves manually transferring pollen from the stamen of one flower to the pistil of another, ensuring genetic control and preventing cross-pollination from unwanted sources. This is especially important for maintaining purity in cross-pollinating species where natural pollination can easily introduce undesired traits. It's a labor-intensive process, but for those preserving rare or highly valued varieties, this level of attention to detail yields invaluable results. While seemingly simple, consistent and careful practice is required to master this crucial step for maintaining seed purity. Without proper technique, even manual pollination can introduce unintended genetic variation.

Roguing, the process of identifying and removing undesirable plants within your seed crop, is a proactive step in preserving purity. Rogue plants might display traits you wish to exclude, such as susceptibility to disease, undesirable fruit size or taste, or an overall weakness in the plant itself. Removing them early prevents their pollen from contaminating your main crop. This prevents future generations from inheriting these negative traits, preserving the desired characteristics of your seeds. Diligent roguing requires keen observation and consistent vigilance throughout the growing season. It's best to begin roguing as early as possible so you don't let negative traits take root.

Proper seed cleaning and storage are the final, but equally important stages in preserving seed purity. Thoroughly cleaning seeds removes unwanted plant matter and other debris which might contain seeds from other varieties or carry contaminants. Seeds should also be stored in a cool, dry place, in airtight containers, to prevent degradation and ensure their viability. Improper storage increases the risk of spoilage, reducing both the quantity and quality of your seed stock and potentially creating ideal conditions for fungal growth or other sources of contamination which can weaken even healthy seeds. This process secures the hard-earned purity you've achieved throughout the growing season and ensures the integrity of your seed stock for future planting.

6 Chapter 6: Policy & Politics

Understanding the political landscape is crucial for anyone serious about sustainable food systems. We can't simply grow food; we need supportive policies to make it truly viable. Think of it like building a house – you need strong foundations, and those foundations are built with the right laws and regulations. Without them, even the most meticulously planned garden will struggle to thrive.

Local food systems often face challenges from larger, industrial agricultural operations. These giants wield significant political influence, often lobbying for policies that favor their large-scale production methods over more sustainable alternatives. This creates an uneven playing field, making it difficult for small farmers and community gardens to compete. We must advocate for policies that level this playing field.

Food sovereignty is a powerful concept, representing the right of people to define their own food and agriculture systems. It's about prioritizing local food production, ensuring fair prices for farmers, and protecting biodiversity. Imagine a community where food is produced and distributed locally, fostering self-sufficiency and minimizing dependence on global supply chains that are often vulnerable to disruptions and inequalities. This requires policies that actively support local farmers and sustainable farming practices.

One way to foster food sovereignty is through direct engagement with local government. Attend town hall meetings, write letters to your elected officials, and support candidates who advocate for sustainable agriculture. Your voice matters, and collective action can create significant change. Support organizations dedicated to food justice and sustainable agriculture. Their expertise and advocacy work amplify your voice and increase the impact of your efforts.

Furthermore, supporting policies that incentivize sustainable practices is critical. This could include tax breaks for farmers who adopt regenerative agriculture methods, subsidies for water conservation technologies, or funding for research into climate-resilient crops. These financial incentives can encourage farmers to transition to more sustainable methods, even if they involve higher initial costs. Remember, the long-term benefits of soil health, water conservation and reduced environmental impact far outweigh the immediate expenses.

Consider the impact of trade agreements. These agreements can significantly impact domestic food production, often favoring imports over locally-grown produce. It's vital to analyze the potential consequences of these agreements and support policies that protect domestic farmers and promote fair trade practices. This might involve actively campaigning for amendments or seeking alternative trade arrangements that prioritize sustainability.

Education plays a vital role in shaping effective policy. We need to raise awareness among the public and policymakers about the importance of sustainable food systems, climate change impacts, and the social and economic benefits of supporting local farmers. Education programs, workshops, and public awareness campaigns can empower individuals to make informed choices and advocate for better policies. This informed citizenry is pivotal for building a strong and resilient food future.

Finally, we must not underestimate the power of collaboration. Successful policy change rarely happens in isolation. Forming alliances with other organizations, farmers' groups, and community activists can amplify your voice and create a stronger advocacy force. Collective action through these collaborations is essential in influencing policy decisions, as unified groups can present a compelling case for change. Remember your role is critical, and your combined actions can lead to transformative change. By working together and advocating for smart, equitable policies, we can create a food system that sustains both humanity and the planet.

6.1 Food Sovereignty

Food sovereignty isn't merely about having enough to eat; it's about controlling the entire food system, from seed to table. It's about reclaiming power over our diets, our livelihoods, and our environment. This means prioritizing local food production, supporting small-scale farmers, and rejecting the globalized industrial food system that often prioritizes profit over people and planet. We must understand that this isn't a utopian ideal but a fundamental right, vital for the health and well-being of communities everywhere.

Shifting towards food sovereignty demands a conscious effort to break free from reliance on large corporations and distant markets. Imagine your community, not reliant on unpredictable global supply chains, but actively engaged in cultivating, processing, and distributing its own food. Picture a system where the decisions regarding your food are made locally, empowering you and your neighbors. This means actively engaging in sustainable agricultural practices, focusing on techniques that restore and nurture local ecosystems.

The transition requires a multifaceted approach, extending far beyond individual choices. It necessitates political action, demanding policies that favor small farmers and sustainable practices. We need to advocate for policies that support local food systems, regulate industrial agriculture's negative impacts, and promote fair trade practices. This isn't about rejecting progress but about redirecting it towards a more just and sustainable path. We must actively participate in shaping agricultural policies that truly reflect the needs of our communities.

Support for local farmers is crucial. By buying directly from them, whether at farmers' markets or through community-supported agriculture (CSA) programs, we strengthen local economies and build more resilient food systems. This directly supports those who cultivate our food, providing them with fair compensation for their labor. Direct engagement, building relationships with these farmers, enhances understanding and fosters a sense of community around food production. Consider the sheer economic power of conscious consumption; with collective action, we can reshape the marketplace to prioritize sustainability.

Seed saving, as detailed in the previous chapter, plays a pivotal role in this journey towards food sovereignty. By saving and exchanging seeds within our communities, we maintain biodiversity, adapt to local conditions, and reduce reliance on corporate seed providers. Consider the generations of knowledge encapsulated within a single seed, representing a direct connection to our agricultural heritage. Seed saving empowers individuals, allowing them to determine the future of their food supply.

Community gardens are tangible expressions of food sovereignty. They provide access to fresh, healthy food, foster social connections, and empower individuals to take control of their food sources. These gardens are not merely places to grow food; they are vibrant spaces that build community, promote healthy eating habits, and offer opportunities for education and skill-sharing. Their impact is felt not just through nourishment, but through community building itself. They are microcosms of the larger food system, demonstrating the feasibility and benefits of community-led initiatives.

Navigating this shift requires education and collaboration. We must learn sustainable farming techniques, understand the politics of food, and work together to build more just and resilient food systems. This involves fostering a deeper understanding of the environmental impacts of food production, focusing on strategies for mitigating climate change and adapting to resource scarcity. It's a collaborative endeavor, requiring education initiatives to teach people of all backgrounds about sustainable farming practices.

Food sovereignty isn't a destination; it's an ongoing process of learning, adapting, and advocating for change. It requires a paradigm shift, moving away from a purely economic view of food production to one that values ecological sustainability and social justice. This shift necessitates understanding the interconnectedness of food, environment, economy, and society. Embracing this holistic perspective is key to achieving true food security and equitable food access.

This journey requires patience and persistence. Building resilient food systems takes time, effort, and collective action. However, the rewards are immense: healthier communities, stronger economies, and a more sustainable future. The transformation towards food sovereignty is a vital step toward safeguarding our collective well-being, ensuring that food isn't merely a commodity but a foundation upon which communities thrive and flourish. It is a commitment to a future where food is a source of nourishment, empowerment, and community strength. It's a future worth fighting for.

6.2 Sustainable Policies

Sustainable agricultural practices are not merely a set of techniques; they are the bedrock of a resilient and equitable food system. Their success hinges on supportive policies that incentivize adoption and address the inherent challenges. Governments and policymakers must recognize this critical role and design interventions accordingly. These policies should not just be reactive, addressing immediate crises, but proactive, anticipating future needs and fostering long-term sustainability.

Direct financial support for farmers transitioning to sustainable practices is paramount. This could manifest as grants, subsidies, or tax breaks specifically targeted at organic farming, agroforestry, and other regenerative techniques. Equally crucial is the establishment of robust certification and labeling systems that guarantee the authenticity of sustainably produced goods, allowing consumers to make informed choices and rewarding farmers for their efforts. These schemes should be transparent, accessible, and rigorous, minimizing the administrative burden on farmers and maximizing their effectiveness. Furthermore, a critical element often overlooked is the provision of easily accessible training and extension services, bridging the gap between knowledge and practical application for farmers.

Beyond direct financial aid, governments play a crucial role in creating a supportive regulatory environment. This includes measures that reduce reliance on synthetic fertilizers and pesticides. Policies promoting the use of organic fertilizers and biological pest control methods are essential. Stringent regulations on the use and disposal of chemical inputs should accompany supportive incentives for sustainable alternatives. Such regulations should be balanced, acknowledging the need for safe and effective pest management while simultaneously prioritizing environmental protection and human health. This also necessitates comprehensive land-use planning that prioritizes agricultural land preservation, minimizing soil degradation and fragmentation. Equally crucial is the preservation of biodiversity, which can be achieved through policies protecting wild pollinators and promoting habitat diversity within and around farmlands.

Investment in research and development is another critical aspect of effective policy. Funding should be directed towards enhancing drought-resistant crop varieties, improving water-use efficiency in irrigation systems, and developing innovative soil health management practices. Such investments are not merely expenses; they are strategic investments that yield long-term benefits in terms of enhanced agricultural productivity, environmental protection, and economic growth. The focus should extend beyond technological advancements, incorporating sociological research to better understand the adoption challenges faced by farmers and tailor policies to ensure greater effectiveness. This also necessitates open communication and collaboration among scientists, policymakers, and farmers, fostering a shared understanding of the challenges and solutions.

Finally, fostering equitable access to resources and markets is crucial for achieving sustainable food systems. This requires addressing the systemic inequalities that often marginalize smallholder farmers, limiting their access to credit, land, and technology. Policies promoting fair trade practices, supporting cooperative farming models, and ensuring equitable land distribution are critical steps towards creating a just and sustainable agricultural sector. It demands not only legislative action but also a cultural shift that values the contributions of small-scale farmers and prioritizes the health of both people and planet. Ultimately, creating a sustainable food system is a multifaceted endeavor that requires a holistic approach, integrating environmental protection, economic viability, and social equity. Sustainable policies, therefore, must be comprehensive, adaptable, and grounded in a deep understanding of the complex interrelationships that shape our food systems. Only through such an approach can we truly cultivate a future where agriculture nourishes both humanity and the planet.

7 Chapter 7: Future of Food

Technological advancements are rapidly reshaping the agricultural landscape, presenting both opportunities and challenges for future food security. Vertical farming, utilizing stacked layers in controlled environments, maximizes space and yields, offering potential solutions for urban food production and minimizing land use. However, the energy consumption of these systems needs careful consideration to ensure their overall sustainability. Hydroponics and aeroponics, which cultivate plants without soil, offer further avenues for efficient resource utilization, potentially reducing water usage and the need for pesticides. Yet, the initial investment costs can be substantial, creating barriers for small-scale farmers. Precision agriculture, employing data analytics and sensors to optimize resource allocation, enhances efficiency and minimizes environmental impact. This approach, however, requires specialized knowledge and technological infrastructure, making it less accessible in certain regions. Genetic engineering holds the promise of creating crops with enhanced resilience to pests, diseases, and climate change. However, ethical considerations surrounding genetically modified organisms (GMOs) require thorough investigation and public discourse before widespread adoption.

Global food security faces immense pressures from a growing population, climate change, and economic inequalities. Feeding a burgeoning global population sustainably necessitates innovative approaches to food production and distribution. Efficient food transportation systems are crucial for reducing waste and ensuring that food reaches those who need it most, particularly in regions facing food insecurity. Reducing food waste along the supply chain, from farm to table, is paramount. Improved storage facilities and more effective packaging can significantly reduce spoilage, saving resources and money. Addressing socioeconomic factors contributing to food insecurity is just as important as improving production methods. Sustainable agricultural practices can be promoted through supportive policies and educational initiatives, empowering communities to achieve greater food sovereignty. Investment in agricultural infrastructure and research is equally critical to developing resilient and productive food systems.

Climate change presents significant threats to global food production, impacting crop yields, water availability, and livestock health. Rising temperatures and changing rainfall patterns can severely diminish crop productivity, especially in regions already vulnerable to food insecurity. Extreme weather events, such as droughts and floods, can disrupt agricultural production and distribution networks. Developing climate-resilient crops through breeding programs and genetic modification is crucial for mitigating the effects of climate change. Implementing sustainable agricultural practices, like no-till farming and agroforestry, can enhance soil health and water retention, improving resilience to climate variability. Diversification of crops and livestock reduces reliance on single species susceptible to climate-related threats. The integration of climate-smart agriculture into national and international policies can support long-term adaptation and mitigation efforts. Investing in early warning systems for climate-related disasters allows for timely interventions to protect food production. Adapting farming practices to local climate conditions is essential for ensuring food security in a changing world.

The future of food requires a multifaceted approach, integrating technological innovation with sustainable practices and equitable policies. The collaborative effort of scientists, policymakers, farmers, and consumers is crucial for shaping a food system that is both productive and sustainable. It demands a shift towards sustainable agricultural practices that minimize environmental impact while maximizing resource efficiency. Embracing diverse food production systems, including urban farming and community gardens, increases resilience and improves accessibility to nutritious food for all. Supporting small-scale farmers and local food systems is vital for fostering economic resilience and community empowerment. Promoting equitable food distribution networks ensures that food reaches those who need it, reducing food insecurity and malnutrition. Strengthening international cooperation on climate change adaptation and mitigation is essential for protecting global food security. Ultimately, the future of food hinges on our collective ability to create a food system that is environmentally responsible, socially just, and economically viable. This means focusing on collaboration, innovation, and a commitment to leaving a sustainable future for generations to come. The transition to this future will require a fundamental shift in our thinking, from a focus on short-term profits to a long-term vision of sustainability and equity. It necessitates the integration of diverse perspectives and approaches. Only through such concerted efforts can we achieve a future where food security is a reality for all.

7.1 Technological Advancements

Precision agriculture employs technologies like GPS, sensors, and data analytics to optimize resource use. Imagine a farmer using a drone to survey their fields, identifying areas needing more water or fertilizer. This targeted approach minimizes waste, boosts yields, and reduces the environmental footprint considerably, a far cry from blanket application methods of the past. Such efficiency is not just about cost savings; it's crucial for sustainable food production, especially given escalating resource scarcity and climate change impacts. This targeted approach also allows for personalized responses to specific plant needs, further enhancing yield and resilience.

Vertical farming stacks crops in layers, maximizing land use in urban environments. Think skyscrapers dedicated to growing food, eliminating the need for vast expanses of arable land. This innovative approach offers potential solutions to food deserts and reduces the transportation distances for fresh produce, ultimately lowering carbon emissions and increasing accessibility to healthy, nutritious food. The controlled environment of these vertical farms also minimizes water waste and pesticide use while providing a consistent yield regardless of external weather conditions. This technology isn't confined to large-scale operations either; smaller, modular systems are increasingly affordable and adaptable for individual use, even on rooftops or balconies.

Robotics and automation are transforming agricultural practices, from planting and harvesting to weeding and pest control. Autonomous tractors and harvesters can work continuously, reducing labor costs and improving efficiency. Automated weeding systems selectively eliminate weeds, reducing the need for herbicides, positively impacting environmental health.

These advancements allow farmers to focus on other critical aspects of their operation, such as planning and quality control, resulting in a more efficient and sustainable agricultural business model. While the initial investment in robotic systems can be substantial, the long-term benefits, in terms of yield, labor efficiency, and reduced environmental impact, are undeniable.
Gene editing technologies, such as CRISPR, offer the potential to develop crops with enhanced traits like drought resistance, pest resistance, and increased nutritional value. Imagine crops that thrive in arid regions or require fewer pesticides, thereby protecting biodiversity and the health of consumers. This targeted approach to crop improvement is significantly faster and more precise than traditional breeding methods. While ethical considerations and regulatory frameworks surrounding gene editing are still developing, the potential to address global food security challenges through this technology is immense.

Improved data management and predictive analytics can help farmers make informed decisions based on real-time data. This involves utilizing weather forecasts, soil analysis, and crop growth models to anticipate potential problems and implement appropriate solutions. This data-driven approach allows for optimized planting schedules, irrigation strategies, and resource allocation, resulting in improved yields and reduced risks. The increasing availability of affordable sensors and data analysis tools makes this technology accessible to farmers of all scales, empowering them to make informed decisions and enhance their resilience in the face of climate variability. Farmers can actively participate in research, offering their own data to contribute to broader agricultural knowledge and innovation.

Blockchain technology can improve traceability and transparency throughout the food supply chain. Imagine being able to trace the origin of your food from farm to table, knowing its exact journey and production methods. This improved transparency can help build trust with consumers, combat food fraud, and ensure fair compensation for farmers. This technology also enables more efficient management of supply chains, streamlining logistics and reducing waste throughout the system. The implementation of blockchain is continuously evolving and its impact in creating greater accountability and integrity within the agricultural sector has the potential to revolutionize the food system.

These technological advancements offer exciting possibilities for enhancing food production, but responsible implementation is crucial. The ethical implications of certain technologies need careful consideration, ensuring that these advancements benefit both the environment and society. Open dialogue, collaboration, and informed policymaking are vital to harness the full potential of these innovations while mitigating potential risks and ensuring equitable access to the benefits. The future of food security hinges on our ability to strategically integrate these technologies into sustainable agricultural practices. The integration of technology requires a holistic view encompassing environmental sustainability, economic viability, and social justice to be successful. It's not merely about technological adoption, but about responsible stewardship of these innovative tools to build a more secure and equitable food system for all.

7.2 Global Food Security

Global food security isn't just about having enough food; it's about ensuring everyone has access to nutritious, safe, and affordable food consistently. This requires a multifaceted approach, transcending national borders and incorporating sustainable practices. We must move beyond reactive measures to proactive strategies that address the root causes of food insecurity. This means understanding complex interactions between climate change, population growth, and economic inequalities. For example, droughts exacerbated by climate change can devastate harvests in vulnerable regions, leading to price spikes and widespread hunger. Similarly, socioeconomic disparities often determine who bears the brunt of these food shortages. Addressing global food security necessitates a commitment to equitable distribution systems and resilient agricultural practices.

The challenge of feeding a growing global population while mitigating the effects of climate change is immense, demanding innovative solutions. Technological advancements, such as precision agriculture and improved crop varieties, offer promising avenues. Precision agriculture, utilizing technologies like GPS and sensors, optimizes resource use, reducing water and fertilizer waste. Simultaneously, developing crop varieties that are resilient to drought, pests, and diseases is crucial for enhancing food production in challenging environments. However, technology alone is insufficient; it must be coupled with responsible agricultural practices and equitable access to resources. Without considering the social and economic contexts, technological advancements can exacerbate existing inequalities.

Sustainable agricultural practices are fundamental to long-term food security. This includes promoting biodiversity, protecting natural resources, and reducing our reliance on synthetic inputs. Diversifying crops, integrating livestock, and practicing agroforestry create more resilient and productive systems, less vulnerable to shocks. For instance, intercropping, a technique of planting different crops together, improves soil health, pest control, and yield compared to monoculture farming. Furthermore, reducing reliance on

chemical fertilizers and pesticides protects ecosystems and human health, promoting environmentally sustainable food production. These strategies, while effective, require significant shifts in farming practices and policies.
Collaboration is key. International cooperation is vital for addressing global food security challenges. This includes sharing knowledge and technologies, supporting developing nations in building resilient agricultural systems, and establishing robust trade agreements that ensure fair prices for farmers and food accessibility for consumers. Developing nations often face unique challenges, including limited access to resources, technology, and infrastructure. International aid and capacity-building programs are essential to empower these nations to enhance their food security. Furthermore, promoting fair trade practices enables farmers in developing countries to receive just compensation for their produce, improving their livelihoods and fostering sustainable agriculture.

Education and awareness are essential components of achieving global food security. Educating farmers about sustainable agricultural practices and empowering consumers to make informed food choices are crucial steps. Promoting dietary diversity and reducing food waste are equally important. Many developing countries face significant post-harvest losses due to inadequate storage and transportation infrastructure. Investing in infrastructure and training programs can significantly reduce these losses, improving food availability. Consumers, through conscious purchasing decisions, can support sustainable farming practices and reduce their environmental impact. Ultimately, achieving global food security is a collective responsibility, requiring collaboration between governments, organizations, farmers, and consumers.

Long-term solutions to global food security demand a systemic change. This requires addressing the root causes of hunger and malnutrition, including poverty, inequality, and conflict. Sustainable development goals provide a framework for achieving these broader objectives. The transition to more sustainable food systems requires policy interventions, infrastructure development, and investment in research and development. Policy changes should incentivize sustainable agricultural practices, support smallholder farmers, and ensure equitable access to resources. Investing in infrastructure, such as irrigation systems and storage facilities, improves agricultural productivity and reduces post-harvest losses. Research and development play a crucial role in developing climate-resilient crop varieties and efficient farming techniques.

Ultimately, the path to global food security is paved with a combination of technological innovation, sustainable practices, collaborative efforts, and policy reforms. It's a journey that demands continuous learning, adaptation, and a shared commitment to ensuring a future where everyone has access to sufficient, nutritious, and safe food. It's about building resilience, not only in our farming systems but in our communities and global partnerships. The journey demands constant vigilance and a commitment to long-term sustainable solutions. The complexities involved require a global effort, one that blends scientific advancements with social and economic considerations. Only then will we truly safeguard the future of food.

7.3 Climate Change Impacts

The escalating impacts of climate change pose a significant threat to global food security, demanding immediate and adaptive strategies within agriculture. Increasing temperatures lead to reduced crop yields and altered growing seasons, impacting the availability and affordability of staple foods worldwide. This is particularly acute in already vulnerable regions, exacerbating existing inequalities and potentially triggering widespread food shortages. We are not just talking about minor inconveniences; we are discussing potential widespread famine and societal collapse.

Heat stress significantly diminishes crop productivity. Many crops, particularly those sensitive to high temperatures, experience reduced flowering, pollination, and fruit set under extreme heat conditions. This leads to smaller harvests, impacting both farmers' livelihoods and consumer access to food. Furthermore, unpredictable rainfall patterns, another hallmark of climate change, make it increasingly difficult to plan planting and harvesting schedules. Farmers face increased uncertainty, leading to potential crop failures and economic hardship. The unpredictability can lead to severe issues, like an inability to properly harvest and store crops due to unanticipated rainfall.

Water scarcity, intensified by climate change, is another critical challenge. Droughts become more frequent and severe, limiting irrigation water availability and stressing crops. This necessitates innovative water management practices, such as rainwater harvesting, efficient irrigation techniques, and the adoption of drought-resistant crop varieties. This isn't simply about saving water; it's about adapting to a drier world, which may mean reevaluating the types of crops we grow. We may need to implement techniques that are unheard of in our current systems.

Pest and disease outbreaks are also projected to increase in frequency and severity with a changing climate. Warmer temperatures and altered rainfall patterns create favorable conditions for many pests and pathogens, leading to greater crop damage and higher reliance on pesticides. This presents challenges for sustainable agriculture, demanding integrated pest management strategies that minimize environmental impact. This also means paying more attention to choosing pest-resistant crops, something that will require extensive experimentation, and potentially, genetic modification.

Sea-level rise poses a direct threat to coastal agricultural lands. Saltwater intrusion contaminates freshwater sources and soil, rendering them unsuitable for crop production. Low-lying coastal communities that heavily rely on agriculture face displacement and loss of livelihoods as fertile land is lost to the rising tides. This represents not just an agricultural crisis, but also a profound humanitarian crisis. Millions could be displaced by these shifts and we must proactively address this in our strategy to ensure food security.

Addressing these impacts necessitates a multi-faceted approach. This involves investing in climate-resilient agricultural practices, including drought-resistant crops, improved irrigation systems, and integrated pest management strategies. Furthermore, robust infrastructure needs to be built to support adaptation to changing climate conditions. These changes must not just be reactive, they should be proactive measures that prevent these issues rather than fixing them as they arise. We need to fundamentally restructure our agricultural policies and practices.

Promoting crop diversification is crucial for building resilience against climate change. A diverse range of crops reduces the risk of widespread crop failure due to climate-related events. This approach also enhances biodiversity, increasing the resilience of entire ecosystems. It is not just about selecting one best crop; it's about having various options that can ensure that even if some fail, others will thrive. This will provide us with a more sustainable and adaptable agricultural system.

Investing in climate-smart agriculture research and development is paramount to develop innovative solutions for climate change adaptation and mitigation. This includes breeding climate-resilient crop varieties and exploring new agricultural techniques that minimize environmental impact and enhance productivity in challenging climates. We must also consider new technologies, such as precision agriculture and vertical farming, that can enhance efficiency and minimize the environmental footprint of agriculture. This is about long-term thinking, and a long-term commitment to research and development.

Strengthening early warning systems for climate-related hazards, such as droughts and floods, enables farmers to make informed decisions and reduce potential losses. These systems should be accessible to all farmers, including smallholder farmers in developing countries, who are particularly vulnerable to climate change impacts. This requires technological infrastructure, but also robust data collection and dissemination systems that reach the people who need it most. It requires a commitment to bridging the information gap between policy makers and producers.

Ultimately, addressing climate change impacts on food security demands a global, coordinated effort involving governments, researchers, farmers, and consumers. This includes implementing policies that incentivize sustainable agricultural practices, ensuring equitable access to resources, and fostering international cooperation to address this complex challenge. This isn't about one country's responsibility; it is about the global community recognizing the shared threat and the shared need to respond. We need a united front in the face of this global crisis. The future of food security depends on it.

www.ingramcontent.com/pod-product-compliance
Lightning Source LLC
LaVergne TN
LVHW022022200926
848481LV00007B/2495
* 9 7 9 8 3 4 9 4 7 1 7 0 4 *